IGNOTUS

The New Jersey Enigma

Decoding the UFO Drone Phenomenon

For inquiries, please contact the author at: ignotus.books@gmail.com.

Audio ISBN: 979-8-9922631-4-5

First edition

ISBN (print): 979-8-9922631-8-3
ISBN (digital): 979-8-9922631-3-8

This book was professionally typeset on Reedsy.
Find out more at reedsy.com

Contents

Common Definitions

1.UFO (Unidentified Flying Object)

A **UFO** refers to any aerial phenomenon that cannot be immediately identified by the observer. While historically associated with extraterrestrial spacecraft, the term simply denotes any object seen in the sky whose nature is unknown at the time of sighting. These objects may later be identified as conventional aircraft, weather phenomena, astronomical objects, or other man-made or natural occurrences.

2.UAP (Unidentified Aerial Phenomena)

In recent years, the term **UAP** has gained popularity, particularly within official government and military circles, as a broader, more neutral term to describe unexplained aerial objects or events. UAPs encompass not just flying objects but also any unusual, unexplained sightings in the atmosphere, regardless of whether they are tangible objects or other atmospheric phenomena.

3.Flying Saucer

A **Flying Saucer** is a specific type of UFO often depicted as a disc-shaped or saucer-like craft. The term became widely popular in the 1940s and 1950s after several UFO sightings and reports featured saucer-like shapes. However, this term has fallen out of favor in more recent discussions, as UFOs are seen in a variety of shapes and forms.

4.Alien Spacecraft

Alien spacecraft refers to the hypothetical idea that UFOs may be vessels piloted by extraterrestrial beings. This concept is often a subject of science fiction and has been speculated by UFO enthusiasts who believe some UFO sightings may involve interplanetary or interstellar travel by beings from other worlds.

5.ET (Extraterrestrial) Craft or Beings

The term **extraterrestrial** refers to any life or craft that originates from outside Earth. **ET crafts** are speculated to be advanced vehicles operated by extraterrestrial beings, which some believe are the source of certain UFO sightings. The idea of **ETs** and their possible interactions with Earth is central to many UFO theories and speculations.

6.Drone

A **drone** refers to an unmanned aerial vehicle (UAV) used for a variety of purposes, including military operations, surveillance, and recreational use. In the context of UFO sightings, drones are sometimes suggested as a potential explanation for certain unexplained aerial phenomena, particularly those observed in urban or populated areas.

7.Military Aircraft

Military aircraft, especially **stealth** or **classified** planes, have often been linked to UFO sightings, particularly when the objects exhibit high-speed maneuvers, unusual shapes, or other characteristics not typically associated with civilian aircraft. Some UFO reports have been attributed to the testing of cutting-edge military technology.

8.Balloon or Weather Phenomena

Some UFO sightings have been explained by meteorological phenomena such as **weather balloons**, atmospheric distortions, or **cloud formations** that can create optical illusions, making objects appear to move or change shape in unusual ways.

9.Interdimensional or Time Travelers

Some theories, particularly those found in speculative science fiction, suggest that UFOs may not be extraterrestrial in origin, but instead are **interdimensional** or from the future. This hypothesis posits that the UFOs could be traveling from parallel universes or even different points in time, explaining their seemingly advanced technology and anomalous behavior.

These definitions help clarify the different terms used in UFO discussions and highlight the variety of interpretations and speculations that surround these mysterious phenomena.

1

Introduction

The skies over New Jersey have long been a canvas for mystery and intrigue. From the infamous *"Great Airship Wave"* of the late 19th century to the countless UFO sightings reported in the modern era, this region has earned a reputation as a hotspot for aerial anomalies. But why New Jersey? Is there something unique about its geography, its people, or its airspace that invites these phenomena?

In recent years, a new player has entered the scene: drones. These advanced, often unmanned aerial vehicles have opened up a new frontier in technology and exploration, but they've also muddied the waters when it comes to identifying what's truly out of the ordinary in our skies. Could the recent sightings of unexplained lights and objects in New Jersey's skies be attributed to advanced drone technology, or is there something even more mysterious at work?

This book seeks to unravel the threads of a complex tapestry, exploring every angle of the New Jersey UFO and drone phenomena. We'll delve into the detailed accounts of eyewitnesses, examine the technological capabilities of drones and experimental aircraft, and consider the possibility of natural phenomena being misinterpreted as something extraordinary. We'll also dive into the cultural and psychological factors that shape how we perceive the unknown.

But no exploration of this topic would be complete without addressing the elephant in the room: the possibility of extraterrestrial involvement. Are we witnessing advanced alien technology? Or are these sightings the result of humanity's ever-expanding technological prowess?

This journey is not about confirming or debunking any one theory. Instead, it's about fostering curiosity, encouraging critical thinking, and embracing the mystery of what we don't yet understand. The New Jersey skies hold secrets that challenge our perception of reality. Together, we'll sift through the evidence, separate fact from fiction, and explore the many dimensions of what might be unfolding above us.

Whether you're a skeptic, a believer, or simply curious, this book invites you to look up, wonder, and question. What's out there? And more importantly, what does it mean for us here on Earth?

2

The Incident - Viral UFO Sightings in New Jersey (2024)

A Detailed Timeline of the Event

In the past few days, **New Jersey** has become the focal point for a series of **UFO sightings** that have garnered widespread attention online. The incidents, which began to unfold on **December 18, 2024**, are marked by **shocking footage, eyewitness reports**, and an **unprecedented level of public curiosity**, quickly spreading across social media platforms.

- **December 18, 2024 - 6:00 PM**: The first significant report came from **Middletown**, New Jersey, where a local resident posted a video on TikTok showing a **bright, pulsating light** moving erratically in the sky. The video, recorded with a smartphone, quickly went viral, amassing thousands of views within minutes. As the light changed colors and zipped across the sky, the footage left many viewers speculating whether it was a drone, a military aircraft, or something else entirely.
- **December 19, 2024 - 7:30 PM**: Another video surfaced from **Hackensack**, New Jersey, showing a formation of **three bright orbs**

in the sky, seemingly rotating in synchronized patterns. The objects appeared to shift direction suddenly, prompting many online viewers to call into question the possibility of these being **military drones** or even extraterrestrial vehicles. Within hours, the video spread across platforms like Twitter, Instagram, and YouTube.

- **December 20, 2024 – 9:00 PM**: A **live stream from Bayonne**, New Jersey, captured multiple unidentified lights hovering in a triangular formation. The stream went viral on both TikTok and Instagram, attracting attention from UFO enthusiasts and news outlets. Many viewers noted how the lights seemed to hover without any visible means of propulsion. The stream was shared by several local news stations, adding credibility to the sightings.

- **December 21, 2024 – 11:00 PM**: A new video surfaced from **Jersey City**, showing a **fast-moving craft** with **multiple lights** darting across the sky. The video, taken by an anonymous source, raised alarms due to its high quality and clear footage. Eyewitnesses claimed they saw the craft moving with speeds and agility far beyond what conventional aircraft can achieve.

As these incidents continued to unfold, a **pattern** emerged: similar **UFO sightings** were reported in multiple towns across New Jersey, with a clear spike in activity around 9:00 PM each evening. Many of these videos were recorded on smartphones, with some users even documenting the sightings live as they happened. The number of reports continued to grow, as more residents came forward with footage of unusual aerial phenomena.

Eyewitness Accounts: Stories from the Ground

The widespread circulation of videos and social media posts allowed

eyewitness accounts to spread quickly, with numerous individuals describing eerily similar phenomena. Several residents shared their experiences, adding to the viral conversation online.

Emily Thompson, Middletown Resident (via TikTok):

"I thought it was just another plane, but then it stopped and hovered for a moment before shooting off at an impossible speed. It was like nothing I've ever seen before. I immediately grabbed my phone and filmed it. When I posted it, the response was insane. People from across New Jersey started saying they saw the same thing!"

Tom Harrison, Hackensack Resident (via Instagram):

"I saw three lights in the sky. They seemed to be in a perfect triangle. They weren't blinking like airplane lights, and they didn't make any noise. They moved in ways I've never seen before. It was hard to believe what I was seeing, so I recorded it. When I shared it, I didn't expect so many others to have captured the same thing."

Rachel Foster, Bayonne Resident (via YouTube live stream):

"I was out with my friends when we saw what looked like multiple lights in the sky. I pulled out my phone and started streaming because I knew this was something big. People started joining my live stream, and we all saw the same thing. It was a massive, bright formation, and it stayed in place for what felt like forever. Then, it just disappeared."

These firsthand accounts continued to fuel online discussions, which only escalated as new videos and eyewitness reports surfaced each night. The convergence of **social media**, **viral videos**, and **eyewitness testimonies** turned the events into a national conversation.

Official Statements: Local Authorities and Government Agencies

As the sightings began to gain widespread attention, **local authorities** and **government agencies** struggled to provide clear answers. The rapid spread of footage and reports only deepened public suspicion and uncertainty.

- **New Jersey State Police**:

"We are aware of multiple reports concerning unusual lights in the sky. Officers are investigating and will continue to monitor the situation. At this time, no official cause has been identified."

- **FBI**:

"The FBI is looking into the recent events surrounding these aerial phenomena. We are working with local law enforcement and other agencies to assess the situation. It is too early to make any definitive statements regarding the nature of these objects."

While both the **State Police** and **FBI** acknowledged the incidents, they offered little in the way of concrete explanations. The lack of a clear and direct response from **government agencies** only fueled further speculation online, with many accusing authorities of withholding information or being ill-prepared to deal with the scale of the phenomenon.

- **Department of Homeland Security (DHS)**:

"We have received numerous inquiries regarding recent aerial sightings. At this time, we are not ruling out the possibility of military or civilian drone activity, but the origins of these sightings remain unknown. We are coordinating with appropriate agencies to investigate these reports."

The **DHS** statement raised even more questions, as it hinted at the possibility of **military drone testing** or **highly classified aerospace technology**, which many conspiracy theorists began to seize upon as potential explanations.

- **Pentagon:**

The Pentagon has addressed the recent surge of mysterious drone sightings over New Jersey, emphasizing that these incidents do not pose a significant threat. In a joint statement with the Federal Bureau of Investigation (FBI), Federal Aviation Administration (FAA), and Department of Homeland Security (DHS), the agencies downplayed the risks associated with the drones, attributing them to various legitimate sources.

President Joe Biden also commented on the situation, asserting that there is "no sense of danger" related to the unexplained drone flights. He noted that the sightings are being closely monitored and involve mostly lawful and benign aircraft activities.

Despite these reassurances, some elected officials, including Senate Majority Leader Chuck Schumer, have called for greater transparency and the use of advanced radar technologies to uncover the origins of the drones. Conspiracy theories have emerged, with some speculating about the involvement of foreign entities or even extraterrestrial sources. However, experts like UFO researcher Nick Pope suggest that the mystery is likely a terrestrial issue, potentially involving foreign adversaries.

- **FAA (Federal Aviation Administration):**

7

In response to the drone sightings, the FAA has implemented a temporary ban on drone flights over critical infrastructure in New Jersey. This ban, effective until January 17, 2025, restricts unauthorized drones from operating within one nautical mile up to 400 feet in 22 areas for security reasons. The FAA has warned of potential deadly force against drones posing security threats and possible detention of drone pilots.

While the Pentagon and other federal agencies continue to monitor the situation, they have not released a specific press release detailing the New Jersey drone incidents. The agencies maintain that the sightings are being investigated, but no immediate threat has been identified.

- **President-elect Donald Trump's Response:**

President-elect Trump has expressed frustration over the unidentified drones and has called for decisive action. He criticized the Biden administration for its lack of response and suggested that the military should shoot down the drones if their origins remain unknown. Trump emphasized the need for transparency, stating that the government knows the source of the drones but has not disclosed this information to the public.

- **Elon Musk's Perspective:**

Elon Musk has taken a different approach, focusing on technological solutions to address the drone issue. He has advocated for the development and deployment of advanced drone detection and countermeasure systems, leveraging his expertise in aerospace technology. Musk's perspective aligns with his broader interest in autonomous systems and their applications in various sectors, including defense.

These differing viewpoints highlight the ongoing debate between immediate military action and technological innovation in addressing the challenges posed by unidentified drones in U.S. airspace.

However, the **lack of transparency** from authorities, coupled with the volume of **viral footage** and **public attention**, left many wondering if there was more to the story than was being shared.

3

Not The First Time

A Detailed Timeline of the Event

On the evening of **August 15, 2024**, New Jersey was thrust into the spotlight as a series of **unexplained aerial phenomena** were reported across multiple towns. This marked the beginning of what would become one of the most widely discussed UFO events in recent history. The timeline below outlines the key moments of this ongoing phenomenon:

- **6:30 PM**: The first reports of unusual lights in the sky were filed in **Toms River**, New Jersey. Multiple witnesses described seeing fast-moving, **drone-like objects** flying in tight formations, with lights flickering in patterns that defied conventional aircraft behavior. These objects moved erratically but silently, leading many to assume they were drones.
- **7:00 PM**: Similar reports began to flood in from **Long Branch**, where residents noted clusters of lights hovering at low altitudes. The objects appeared to move in and out of clouds, sometimes vanishing only to reappear moments later. Eyewitnesses were confused and alarmed, especially given the objects' apparent ability to change

direction quickly.

- **8:00 PM**: A more significant wave of sightings occurred near **Atlantic City**. Witnesses reported seeing several large, **triangular craft** with bright, pulsating lights. These were described as having a slow, deliberate flight pattern. Some witnesses noted that the objects appeared to "pulse" before accelerating at speeds that were impossible for any known human-made vehicle.

- **9:00 PM**: Local police received numerous calls, and authorities dispatched officers to investigate. The **FBI** and **Department of Homeland Security** were alerted, and an emergency meeting was convened. The sightings persisted into the early hours of the morning.

- **1:00 AM** (August 16, 2024): An emergency press conference was held by local authorities. Law enforcement confirmed the sightings but were cautious in their statements, emphasizing that no official identification had been made. They suggested that the objects could be related to military exercises or high-tech drones, but no explanation was given.

- **4:00 AM**: Some of the objects were observed to change formation, at times splitting apart or merging again, making it clear that the event was unlike any typical aircraft sighting. One eyewitness even described an object hovering for several minutes, then shooting straight up into the sky at a remarkable speed.

By the morning of August 16, the media had picked up the story, and the phenomenon had gained national attention. What had started as a local mystery was now being investigated by **federal agencies**.

Eyewitness Accounts: Stories from the Ground

The eyewitness accounts from **New Jersey residents** have been crucial in piecing together the details of the event. These personal stories provide a sense of the widespread nature of the phenomenon and the impact it had on the community.

Rebecca Johnson, Toms River Resident:

"I was sitting on my porch when I first noticed the lights. At first, I thought it was a helicopter, but it was moving in ways that didn't make sense. It stopped, then started moving again, almost like it was following something. I counted at least five objects, all in a perfect line, but they weren't making any noise. That's when I realized something wasn't right."

Tom Harris, Atlantic City Resident:

"The things I saw didn't look like anything I've ever seen before. They were big—triangular—and their lights changed colors. They were so low to the ground, I thought they were going to land, but they didn't. They just hovered there for a while, then darted off at crazy speeds. I've seen drones before, but nothing like this."

Lily Gomez, Long Branch Resident:

"It was like watching a light show, but one that wasn't meant to be seen. These objects seemed to be in sync with each other, moving around in ways that shouldn't have been possible for anything we know of. I tried to get a picture, but my phone wouldn't work when I aimed it at them. It was like something was blocking it."

These accounts, though differing in details, all share common themes: **unusual movement patterns**, **lack of noise**, and the **strange lighting** of the objects. Many witnesses were left with a sense of awe and confusion, struggling to identify what they had seen.

Official Statements: Local Authorities and Government Agencies

As the reports continued to pour in, local authorities were quick to respond. However, the official statements made by law enforcement and government agencies only added to the mystery.

New Jersey State Police:

In the hours following the initial reports, the **New Jersey State Police** issued a statement confirming that officers had been dispatched to investigate the phenomena. While they acknowledged the numerous reports from citizens, they declined to make any definitive conclusions. The police suggested the possibility of high-tech drones being used in some sort of **military exercise**, though they admitted there was no clear explanation for the objects' movement.

FBI:

The **Federal Bureau of Investigation** became involved shortly after the sightings began. FBI spokesperson **Mark Donovan** commented during a televised press conference:

"At this time, the FBI is looking into the reports as part of a broader investigation into what might be classified as aerial phenomena. We urge the public to remain calm and report any sightings to the appropriate authorities. However, we cannot provide further information at this time."

This cautious approach did little to quell growing speculation, with some witnesses suggesting that the lack of clarity only reinforced the notion of a **cover-up**.

Department of Homeland Security:

The **Department of Homeland Security (DHS)** released a more cryptic statement, suggesting that the aerial phenomena could be related to foreign surveillance or technological advancements from other nations. This was a particularly concerning comment, given the rising tensions globally over issues like **nuclear power, military presence**, and **cyber-security**. However, the DHS did not provide any conclusive evidence to support this theory.

Pentagon:

The **Pentagon** also issued a brief statement through the **Department of Defense**:

"We are aware of the reported incidents in New Jersey and are investigating them. The Department of Defense has not confirmed any involvement with the objects observed, but we are looking into all possibilities."

The Pentagon's statement mirrored the cautious tone taken by local authorities. Many UFO researchers and advocates found this response unsatisfactory, claiming that the government was deliberately withholding more information about what had been observed in the skies.

As the incident developed over the following days, speculation grew not only among the general public but also within government circles. Was this an isolated event, or was it part of a larger pattern? Had something, or someone, intentionally made itself visible, or was it just an accident?

4

The Truth About Government Misinformation: The Lost Archives Episode 14 - Unknown Encounter

In the world of UFO research, one of the most significant revelations in recent years has come from the **Lost Archives** podcast series, specifically *Episode 14: Unknown Encounter.* This episode dives deep into the accounts and evidence suggesting that the government, particularly in the United States, has been deliberately misleading the public about UFO sightings and encounters for decades.

The Radio Announcement

In this episode, the Lost Archives team unearthed a radio announcement made by a high-ranking government official, which, when analyzed, contradicted previously stated facts about UFOs and extraterrestrial life. The radio transmission, recorded in the early 1970s, revealed startling information that had been kept under wraps by various government agencies. According to the broadcast, the government had knowledge of unidentified flying objects that were far beyond the technology known to humanity at the time.

This announcement suggested that these craft were not merely misidentified military or weather phenomena, but instead represented something **otherwardly**—potentially extraterrestrial. However, following the broadcast, officials quickly moved to suppress this information, denying any connection between UFOs and alien technology. Over the years, various branches of the government continued to provide misleading, incomplete, or entirely false information to the public, including claiming that UFOs were simply hoaxes or the result of mass hysteria.

The Role of Disinformation

The *Unknown Encounter* episode reveals how disinformation campaigns were intentionally orchestrated to confuse the public and detract from the government's own involvement in UFO investigations. After the 1970s broadcast, where the possibility of extraterrestrial origins was openly discussed, the public narrative shifted dramatically. The government began to introduce conflicting stories—ranging from claims that UFO sightings were simply "weather balloons" to labeling eyewitness accounts as the result of mass psychological phenomena.

As the years went on, classified projects like **Project Blue Book** became key tools in maintaining the government's control over information. Although it officially ended in 1969, many suspect that similar covert programs continued under different names, all designed to keep the UFO phenomena hidden from public scrutiny.

The Alleged Cover-Up

One of the most startling aspects of the episode is the claim that numerous UFO sightings—specifically in regions like New Jersey—were systematically downplayed or dismissed by government officials. Even when high-ranking military and civilian personnel came forward with credible reports, the government maintained a narrative of denial, often through press releases, false public statements, or threats of silence. This strategic obfuscation led to decades of confusion and frustration for both witnesses and researchers.

The *Lost Archives Episode 14: Unknown Encounter* doesn't just highlight how the government covered up the truth about UFOs but also suggests that certain incidents were actively manipulated. The episode includes testimony from former intelligence officers and analysts who claim to have seen firsthand evidence of such a cover-up—suggesting that sightings of UFOs were, in fact, **well-documented by the government**, but suppressed for reasons that are still unclear.

Why This Is Relevant Now

In the context of New Jersey's recent UFO activity, these revelations are profoundly significant. If the government was actively misleading the public about UFOs for decades, it raises questions about the **current wave of UFO sightings** in the state. Are we now seeing a new phase of disclosure, or is the government continuing to withhold information? The connection between past and present UFO encounters suggests that New Jersey may be at the center of a much larger, ongoing mystery— a mystery that government agencies have long sought to control and suppress.

As the public becomes more attuned to the truth about UFOs, events like those described in *The Lost Archives* serve as powerful reminders that the truth is often far stranger than we are led to believe. What we are witnessing today may only be the beginning of a much larger revelation about the nature of UFOs, extraterrestrial life, and the role of government agencies in keeping these phenomena under wraps.

5

Coincidences and Connections: The New Jersey UFO (2014)

In 2014, a book titled *The New Jersey UFO* was published, shedding light on UFO sightings and encounters reported in the state over several decades. The book has drawn attention not only for its insights into the history of UFO sightings in New Jersey but also for the coincidental themes and patterns that appear to align with more recent events, including the wave of sightings in the state in 2024.

Key Themes and Coincidences

1.**Similar Patterns of Sightings**

In *The New Jersey UFO*, the author documented a history of UFO sightings in various towns across the state, ranging from military-style craft sightings to more mysterious, unexplained phenomena. What's striking about these reports is the similarity in patterns—particularly the way in which sightings are often clustered within specific regions or cities, including those around the Jersey Shore and inland areas like

Princeton and Toms River. In recent years, particularly 2024, sightings appear to be repeating these same regional patterns. Could it be that the same areas are experiencing recurring phenomena, or is it merely coincidence?

2.Descriptions of UFOs

In the 2014 book, many witnesses described seeing **unusual, drone-like objects** in the sky—small, fast-moving craft that didn't match conventional aircraft specifications. These descriptions echo the more recent sightings, where reports of similar drone-like UFOs have surfaced. The connection between these two periods raises intriguing questions about whether this type of UFO has been observed consistently over time or if the phenomenon is evolving.

3.UFO Activity in New Jersey as a Hotspot

The 2014 book also identified New Jersey as a hotspot for UFO activity, a theme that continues to hold relevance today. While UFO sightings are reported across the U.S., New Jersey's proximity to major cities like New York and Philadelphia, as well as its rich military history, has made it a focal point for theories surrounding extraterrestrial activity, government involvement, and secret testing. The resurgence of activity in 2024 may be indicative of something much larger at play.

4.Government and Military Interest

The 2014 book also explored possible government involvement or interest in UFO sightings in the region, citing historical documents and reports from military personnel who had witnessed unexplained aerial phenomena. In recent years, with the declassification of certain military UFO encounters, many of the themes discussed in *The New Jersey UFO* have gained renewed attention. Could there be a larger government narrative unfolding that is tied to New Jersey's UFO history?

5.Cultural Influence and Public Perception

Another fascinating coincidence is the way the cultural perception of UFOs and aliens has evolved since 2014. The 2014 book highlighted a shift in public attitudes, where more and more people began to embrace the idea that UFO sightings might be related to something otherworldly rather than simple misidentifications of conventional aircraft. This cultural shift has only accelerated in recent years, with more people open to the possibility of extraterrestrial life and the government's role in keeping information about UFOs secret. The connection between these two time periods speaks to how the conversation surrounding UFOs has grown in recent years, coinciding with an uptick in sightings, particularly in New Jersey.

Why Does This Matter?

The connection between *The New Jersey UFO* and the recent wave of sightings could be more than just a coincidence. The recurring patterns in both the sightings and the public response suggest that New Jersey may continue to be a focal point in the study of UFO phenomena. The fact that the descriptions and patterns observed in 2014 align so closely with recent events adds a layer of intrigue to the situation—one that could indicate something larger at play. Whether it's a reinvigorated interest in UFOs or an actual uptick in activity, the coincidence between the two periods demands attention and careful investigation.

6

Technological Perspectives

In this chapter, we will explore possible technological explanations for the UFO and drone sightings in New Jersey, shifting the focus from extraterrestrial and psychological factors to more grounded, human-made phenomena. With the rapid evolution of technology in the 21st century, the line between what is considered a UFO and what can be attributed to advanced human technology is increasingly blurred. Could the objects observed in New Jersey be cutting-edge drones, experimental aircraft, or something else created on Earth rather than from beyond the stars?

The Rise of Autonomous Drones

One of the most common explanations for modern UFO sightings is the presence of advanced drones. Drones today are capable of flying silently and can perform complex maneuvers, much like the erratic flight patterns often reported in UFO encounters. Some key aspects to consider are:

- **Commercial and Military Drones**: The rapid development of both consumer-grade and military drones has led to an increase in sightings of fast-moving, hovering objects. These unmanned aerial vehicles (UAVs) come in various sizes, from small personal drones to larger military-grade models.

- **Stealth Technology**: Some UAVs employ stealth technology, making them virtually undetectable to radar or the naked eye, adding to their "unidentified" nature.
- **AI and Autonomous Flight**: With AI advancements, drones can now perform autonomous flights, navigating and avoiding obstacles without human intervention. This makes them capable of unpredictable and precise movements that can appear otherworldly.

In New Jersey, reports of drones seen flying erratically or remaining stationary in the sky could potentially be attributed to experimental or military drones operating in the area.

Military and Government Test Flights

Another possible technological explanation is that the sightings could be related to military or government testing. Many military organizations, including those in the United States, are known to test new and classified aircraft. These aircraft, often experimental, may exhibit unusual behaviors or advanced technology that is not yet publicly understood. Some considerations include:

- **Black Budget Projects**: The U.S. government's "black projects" —secretive military programs with undisclosed funding—are known to develop experimental aircraft and technologies. These can range from advanced stealth bombers to hyper-sonic planes.
- **Flight Test Areas**: While New Jersey is not traditionally known for

major military testing grounds, the proximity of large urban centers, including New York City and Washington D.C., may make the region a place for limited, less-publicized test flights. The low visibility and rapid development of these craft could explain why many sightings seem to defy conventional understanding.

- **Test-Flight Operations Gone Awry**: Occasionally, test flights or experimental aircraft can go off-script, leading to unexpected sightings by the public. The reported "uncontrolled" movements and sudden accelerations could very well be related to testing of new aerospace technologies.

Given New Jersey's proximity to key government and military locations, such as the Naval Weapons Station Earle and Picatinny Arsenal, these sightings could be the result of military tests, either deliberately or accidentally exposed to the public.

Hyper sonic and Advanced Aircraft

The cutting-edge capabilities of hyper-sonic flight technologies may also account for some of the UFO sightings. Hyper sonic aircraft can fly at speeds greater than five times the speed of sound (Mach 5), far beyond the capabilities of current commercial or private planes. Key features of hyper-sonic flight include:

- **Extreme Speeds and Maneuverability**: Hyper sonic aircraft can reach incredible speeds, making sharp turns or sudden altitude changes that would be difficult for traditional aircraft to match. The same characteristics have been described in some UFO sightings.
- **Energy Efficiency**: Hyper sonic technology often utilizes new propulsion systems, which can allow for efficient, rapid travel

without the large fuel consumption typically seen in conventional aircraft.

Reports of fast-moving objects with little to no sound could possibly be explained by these advanced, near-orbital aircraft. Could what many have assumed to be UFOs be classified military aircraft undergoing test flights in restricted airspace?

Quantum and Electromagnetic Propulsion

One of the most fascinating and still-theoretical areas of aerospace technology is the concept of quantum or electromagnetic propulsion systems. These systems, which theoretically could harness the forces of electromagnetism or manipulate space-time itself, might explain some of the unexplained elements seen in UFO reports, such as:

- **Anti-gravity Technology**: Some UFOs are reported to hover or move in ways that defy conventional physics, such as remaining stationary in the air and then rapidly accelerating in a different direction. This could be the result of anti-gravity propulsion, a concept studied by physicists but never fully realized in practice.
- **Warp Drives**: Theoretical propulsion technologies like those suggested in quantum physics, such as the Alcubierre Drive, could allow objects to travel faster than the speed of light by bending space-time. These types of technologies, if realized, would explain the sudden accelerations and deceleration observed in UFO sightings.

While these technologies remain speculative, some researchers suggest that what we see in the sky could be the result of breakthroughs that are either highly classified or still in development.

The Role of Atmospheric Phenomena

While drones and military aircraft account for many UFO sightings, there are also natural phenomena that can contribute to misidentifications. Atmospheric and environmental factors could create optical illusions or unusual visual effects that could explain some of the New Jersey sightings, such as:

- **Ball Lightning**: This rare, unexplained phenomenon involves spherical electrical discharges that can hover and move erratically. Some people who witness ball lightning often describe it as a floating, glowing object—characteristics commonly associated with UFOs.
- **Meteorological Anomalies**: Certain weather conditions, such as clouds, storms, or temperature inversions, can create visual effects that appear to distort or "bend" light, making objects in the sky seem to behave strangely.
- **Optical Phenomena**: Mirages, halos, and other atmospheric effects caused by the refraction of light through the atmosphere could explain the appearance of mysterious lights in the sky.

Although less likely to be the cause of the New Jersey sightings, these natural phenomena serve as a reminder that not every unexplained sighting is the result of technology or extraterrestrial involvement.

The Future of UAVs and Aerospace Technology

Looking ahead, we can expect even more technological advancements in the fields of drones, aviation, and aerospace propulsion. Key areas of development include:

- **Urban Air Mobility (UAM)**: As drone technology advances, the potential for widespread use of UAVs for both cargo and passenger transport increases. Could these craft eventually blend into the skyline, confusing the public?
- **Unmanned Aerial Systems (UAS)**: The military and commercial sectors are investing heavily in next-generation UAS, which will feature advanced flight capabilities, stealth, and autonomy. These systems may soon become indistinguishable from UFOs to the naked eye.
- **Civilian Access to Advanced Technology**: As UAVs become more affordable and accessible to civilians, it's likely that more individuals will engage in aerial experiments, creating an environment ripe for new UFO reports.

In New Jersey and beyond, the future of aerial technology will continue to blur the lines between what is considered ordinary and what is truly unexplained. Could the next wave of UFO sightings be nothing more than a preview of the technologies to come?

Conclusion

While extraterrestrial explanations continue to captivate the public imagination, technological factors are increasingly providing plausible alternatives for the UFO sightings observed in New Jersey. Whether it's advanced drones, experimental military aircraft, or yet-to-be-realized propulsion systems, the technological landscape is rapidly evolving in ways that could account for many of the sightings in the area. As technology advances, we may find that many of the "unidentified" objects in our skies have perfectly logical, human-made explanations. But as always, the question remains: is everything as simple as it seems, or could the true answer lie in the unexpected?

7

Extraterrestrial Theories

Few explanations capture the imagination as powerfully as the idea of extraterrestrial involvement. UFO sightings around the world have long fueled speculation about alien visitors observing, studying, or even interacting with humanity. Could the unexplained lights and objects seen in New Jersey's skies represent advanced technology from another world?

This chapter explores the extraterrestrial hypothesis, examining its historical context, the technological implications of alien crafts, and the theories surrounding their potential purposes.

The UFO Phenomenon: A Historical Context

Unidentified Flying Objects have been reported for centuries, with descriptions often transcending cultural and technological boundaries. Key moments in the history of UFO phenomena include:

- **The 1947 Roswell Incident**: A crash in New Mexico that sparked modern UFO lore and claims of government cover-ups.

- **The 1952 Washington D.C. Sightings**: A series of radar and visual observations over the U.S. capital, raising concerns about national security.
- **The Belgian Wave (1989–1990)**: Triangular UFOs observed by thousands, including military personnel.

In the context of New Jersey, several notable UFO sightings and events have occurred, reinforcing its status as a hotspot for aerial mysteries.

Alien Technology: Theories of Surveillance or Exploration

If the objects observed in New Jersey are extraterrestrial, what might they represent?

- **Surveillance Drones**: Alien civilizations could deploy autonomous probes to monitor Earth from a safe distance.
- **Exploration Vessels**: Advanced spacecraft might be used for exploration, much like humans send rovers to other planets.
- **Energy-Based Propulsion Systems**: Observed movements, such as rapid accelerations or sharp turns, could suggest propulsion methods beyond our current understanding, including:
- **Anti gravity Technology**
- **Warp Drives**
- **Electromagnetic Field Manipulation**

Could the UFOs reported in New Jersey demonstrate technological capabilities far beyond what humanity has achieved?

UFOs and Drones: A Connection?

29

Some theorists suggest that what we interpret as drones may actually be alien crafts designed to blend into human technological advancements. Consider the following:

- **Camouflage Technology**: Advanced civilizations may intentionally mimic human-made drones to avoid detection.
- **Reverse Engineering**: Could the rapid rise of drone technology stem from the study of recovered alien materials?

Why New Jersey?

The choice of location for extraterrestrial activity is often debated. In the case of New Jersey:

- **Proximity to Major Cities**: Alien visitors might prioritize observing densely populated areas like New York City.
- **Military Installations**: The region's airbases and research facilities may attract attention from advanced observers.
- **Geographic and Atmospheric Conditions**: Unique environmental factors could make the area suitable for alien monitoring.

UFOs as a Universal Phenomenon

The New Jersey sightings are not isolated. Around the world, similar incidents share common characteristics:

- **Bright Lights and Orbs**
- **Erratic Flight Paths**
- **Silent Movement**

These shared traits suggest a global pattern that transcends cultural and geographic boundaries, strengthening the argument for extraterrestrial involvement.

Skepticism and Scientific Inquiry

Despite the allure of alien theories, many scientists urge caution, advocating for:

- **Rigorous Evidence**: Concrete proof, such as physical artifacts or undeniable video footage.
- **Alternative Explanations**: The application of Occam's Razor to favor simpler explanations.

The U.S. government has recently declassified UFO-related documents through programs like the Pentagon's *Unidentified Aerial Phenomena (UAP) Task Force*, adding credibility to the investigation of these sightings.

Are We Ready for Contact?

The possibility of alien visitors raises profound questions:

- **What Are Their Intentions?**: Observation, contact, or something more ominous?
- **How Would Humanity React?**: Societal, political, and religious implications.
- **Are We Alone?**: What alien life might mean for humanity's understanding of the universe.

The New Jersey sightings could represent a piece of a much larger cosmic puzzle—one that challenges humanity to confront the unknown with an open mind and a spirit of inquiry.

8

Psychological and Sociocultural Dimensions

UFO sightings, including those in New Jersey, often elicit strong emotional responses and foster widespread speculation. While many accounts are rooted in genuine experiences, psychological and sociocultural factors can heavily influence how such phenomena are perceived and interpreted. This chapter delves into the human mind's role in UFO sightings and the cultural forces that shape our collective fascination with the unknown.

The Psychology of UFO Sightings

Our perception of strange phenomena is often shaped by cognitive and emotional processes. Some psychological factors that may contribute to UFO sightings include:

- **Pattern Recognition**: The human brain is wired to identify patterns, even in ambiguous stimuli, such as lights in the sky or distant objects.
- **Pareidolia**: A tendency to perceive meaningful shapes or structures in random visual data (e.g., mistaking a cloud for a flying saucer).

- **Confirmation Bias**: People are more likely to interpret ambiguous events as UFOs if they already believe in extraterrestrial visitation.
- **Stress and Anxiety**: High-stress environments can heighten sensitivity to unusual stimuli, especially in uncertain times.

Could the New Jersey sightings be influenced by psychological factors, particularly in a region known for dense populations and bustling activity?

Mass Hysteria and Shared Experiences

UFO sightings often occur in clusters, with multiple people reporting similar phenomena. This can sometimes result from:

- **Mass Hysteria**: A psychological phenomenon in which a group of people simultaneously experience similar symptoms or perceptions due to collective stress or suggestion.
- **Social Contagion**: Witnesses may unconsciously influence one another, amplifying or reinforcing reports of UFO activity.
- **Media Amplification**: News coverage, social media posts, and online discussions can spread reports rapidly, shaping how events are interpreted.

In New Jersey, the rapid dissemination of sightings through social media may have played a role in shaping public perceptions.

Cultural Narratives and the UFO Phenomenon

Our interpretations of unexplained phenomena are often influenced by cultural narratives. Some key influences include:

- **Science Fiction Media**: Popular films, books, and TV shows often depict UFOs and aliens, shaping expectations about what extraterrestrial encounters might look like.
- **Folklore and Mythology**: Stories of strange lights and flying objects have existed for centuries, rooted in cultural lore and religious traditions.
- **Cold War Paranoia**: The mid-20th century saw a surge in UFO sightings linked to fears of espionage and technological superiority.

Could the modern obsession with UFOs be a reflection of deeper societal anxieties or aspirations?

Witness Credibility and Variability

Not all UFO sightings can be dismissed as psychological or cultural phenomena. Witness accounts vary widely in terms of:

- **Professional Backgrounds**: Reports from pilots, military personnel, and scientists are often considered more credible.
- **Consistency**: The details of some sightings remain remarkably consistent across time and witnesses.
- **Emotional Responses**: Genuine fear, awe, or confusion often accompanies these experiences, suggesting sincerity in many accounts.

New Jersey's diverse population provides a rich tapestry of witness accounts, adding complexity to the narrative.

The Role of Modern Technology

Modern technology has transformed how we document and share UFO sightings. Key developments include:

- **Smartphone Cameras**: Ubiquitous devices allow for quick documentation of sightings, though the quality is often insufficient for definitive analysis.
- **Social Media**: Platforms like Twitter and Facebook facilitate the rapid spread of information (and misinformation) about sightings.
- **AI Algorithms**: Machine learning is being used to analyze patterns in UFO reports, potentially identifying trends or debunking hoaxes.

However, technology also introduces challenges, such as:

- **Deepfakes and CGI**: The rise of digitally altered images and videos has made it increasingly difficult to distinguish genuine evidence from hoaxes.
- **Data Overload**: The sheer volume of reports can overwhelm researchers, complicating efforts to verify claims.

Collective Fascination with the Unknown

At its core, the UFO phenomenon taps into humanity's innate curiosity and desire to explore the unknown. Whether driven by:

- **Fear of the Unfamiliar**: Anxiety about external threats.
- **Hope for Discovery**: The dream of contact with advanced civilizations.

- **Existential Questions**: A yearning to understand humanity's place in the universe.

The sightings in New Jersey, like so many others, are more than just unexplained events. They are reflections of the human psyche, a blend of perception, imagination, and cultural influence.

9

The Alien Hypothesis and Extraterrestrial Encounters

Though technological explanations have grown in credibility, the alien hypothesis remains one of the most captivating and enduring explanations for UFO sightings. The idea that visitors from other planets, galaxies, or dimensions are observing or interacting with Earth continues to be a central element in UFO lore, and it forms the basis of many UFO reports, including those from New Jersey. This chapter explores the possibility of extraterrestrial involvement, the arguments for and against such encounters, and what the implications would be if we were to truly encounter beings from beyond our planet.

The Case for Extraterrestrial Life

The scientific search for extraterrestrial life is an ongoing effort. With the discovery of thousands of exoplanets in the habitable zone of other stars and the growing understanding of extremophiles (organisms that can thrive in extreme environments), the possibility that life exists beyond Earth seems increasingly plausible. Key factors include:

- **The Drake Equation**: A formula developed to estimate the number of active, communicative extraterrestrial civilizations in the Milky Way galaxy. While speculative, it suggests that the probability of alien life is high, especially given the vast number of stars and planets in our galaxy.
- **Recent Discoveries in Astrobiology**: The detection of microbial life, or evidence of the conditions necessary for life, on planets such as Mars and moons like Europa and Enceladus strengthens the hypothesis that life could exist elsewhere in the universe.
- **The Fermi Paradox**: The paradox that despite the vastness of the universe and the high probability of extraterrestrial life, we have yet to find definitive evidence of alien civilizations. This paradox has led some to suggest that alien species are avoiding contact, or that their presence is more subtle than expected.

Could the UFO sightings in New Jersey be evidence that extraterrestrial life has discovered Earth, and is choosing to observe or make contact with us in ways we may not fully comprehend?

UFO Sightings and Alien Encounters

For many, the most compelling UFO sightings are those that involve alleged interactions with extraterrestrial beings. From classic "flying saucer" encounters to more modern sightings of strange, fast-moving objects in the sky, some reports include elements that suggest extraterrestrial involvement. Key aspects of such encounters often include:

- **Close Encounters of the First Kind**: Sightings of lights or objects in the sky that are relatively close to the observer. These are often the most common reports and can be misidentified as aircraft, drones, or atmospheric phenomena.

- **Close Encounters of the Second Kind**: Incidents in which a UFO leaves physical evidence, such as radiation burns, indentations in the ground, or disrupted electrical equipment.
- **Close Encounters of the Third Kind**: The most sensationalized encounters, in which individuals report seeing beings or craft of extraterrestrial origin. These sightings may involve alleged abductions, direct communication, or other forms of interaction.

While many UFO reports can be explained by natural or human-made phenomena, there remain a number of cases that defy easy explanation. Could the New Jersey sightings be part of a larger pattern of extraterrestrial observation or interaction?

The Types of Alien Encounters

If we assume that some UFO sightings are genuinely linked to extraterrestrial life, then the nature of these encounters becomes crucial. Different reports suggest a variety of possible interactions, each with its own implications:

- **Observational Encounters**: Many UFO sightings involve objects that appear to be simply observing Earth from a distance. These encounters may not involve direct contact but could suggest that extraterrestrial civilizations are monitoring humanity's development.
- **Abduction Phenomena**: Perhaps the most controversial aspect of UFO lore is the abduction experience, in which individuals claim to have been taken aboard a spacecraft by aliens for medical or scientific experiments. While many of these reports are dismissed as psychological phenomena or sleep paralysis, some individuals

describe highly detailed, consistent experiences.

- **Contact with Advanced Civilizations**: Some individuals claim to have had direct communication with extraterrestrial beings, either through telepathy, physical meetings, or advanced technology. These experiences often involve messages about peace, the environment, or the future of humanity.

The New Jersey sightings, while not necessarily involving abduction scenarios, may still reflect the possibility of extraterrestrial observation or contact, particularly in light of the sophisticated technology that could be at play.

Alien Technology and UFO Craft

One of the primary reasons that UFOs are often attributed to extraterrestrial life is the advanced technology that some objects appear to demonstrate. Common characteristics of UFO craft include:

- **Non-Newtonian Flight**: UFOs often exhibit flight characteristics that defy the laws of physics as we currently understand them, such as abrupt accelerations, rapid deceleration, and hovering without visible means of propulsion.
- **Silent Operation**: Many UFOs are reported to make no noise at all, which is particularly baffling when considering the lack of traditional propulsion systems.
- **Energy Signatures**: Some sightings report unusual electromagnetic or energy signatures, such as disturbances in electronic devices or powerful bursts of light. These could indicate highly advanced propulsion systems that harness energy in ways not yet understood by human science.

If the New Jersey UFO sightings involve craft demonstrating these capabilities, it may suggest a technology far beyond what humanity currently possesses, supporting the idea of an extraterrestrial origin.

The "Zeta Reticuli" Hypothesis and Grey Aliens

The most well-known depiction of extraterrestrial beings in UFO lore is the "Grey" alien. These small, grey-skinned beings with large heads and black eyes are frequently associated with abductions and UFO sightings. One popular theory, known as the **Zeta Reticuli hypothesis**, suggests that these beings are from the Zeta Reticuli star system. This hypothesis gained widespread attention after the 1961 Barney and Betty Hill abduction, where the couple reported being taken aboard a UFO by beings resembling the classic "Grey" aliens.

- **The Zeta Reticuli Star System**: The idea that alien life could originate from specific stars like Zeta Reticuli or other nearby systems has been a key element in UFO research. Some argue that aliens may be visiting Earth as part of a long-term scientific mission or exploratory endeavor.
- **The Grey Alien Archetype**: This archetype is now entrenched in popular culture, often depicted as a race of beings with advanced technology and an interest in human genetic material. Could this depiction be grounded in reality, and might it explain the nature of certain UFO sightings?

If the New Jersey sightings include reports of humanoid beings or craft resembling those described in alien abduction scenarios, it would add weight to the theory that extraterrestrial entities are visiting Earth.

Government and Military Conspiracy Theories

Some UFO researchers argue that the government has been concealing evidence of extraterrestrial encounters for decades. These theories are fueled by incidents such as the **Roswell UFO Incident**, the release of military footage showing strange aerial phenomena (e.g., the Pentagon's UFO videos), and leaked government documents hinting at knowledge of alien life. These theories suggest that extraterrestrial encounters have been covered up to avoid public panic, military disadvantage, or disruption to societal norms.

- **The "Cover-Up" Theory**: Governments may have been aware of extraterrestrial visitation for years but have withheld this information for various reasons, including national security concerns and the potential for mass hysteria.
- **Whistle-blowers and Leaked Documents**: There are numerous whistle-blowers who claim to have worked on classified projects involving extraterrestrial technology or alien beings. These individuals often point to physical evidence, such as crash sites or recovered materials, as proof of extraterrestrial contact.

In New Jersey, reports of unusual sightings could be seen as part of a larger, secretive narrative. Could the government or military have knowledge of extraterrestrial activity, and is it possible that the public is being kept in the dark?

Conclusion: Extraterrestrial Hypotheses and the New Jersey Sightings

As with all UFO sightings, the question of whether extraterrestrial life is responsible for the phenomena seen in New Jersey remains open. While technological explanations and psychological factors may account for many sightings, there remains a subset of cases that continue to defy logical explanation. The possibility that intelligent life from beyond our planet is visiting or observing Earth adds a layer of complexity to the UFO phenomenon, one that is as much about the mysteries of the universe as it is about our place within it.

Could the New Jersey sightings be a part of a broader, ongoing encounter with extraterrestrial intelligence? Or are we simply witnessing the limits of human understanding, where technology, psychology, and the unknown intersect? As we continue to search the skies, the answer remains elusive, but the search itself may lead us to new discoveries about both the universe and ourselves.

10

Government Secrecy and Military Operations: The Role of Cover-ups in UFO Sightings

Throughout history, governments and military organizations have been deeply involved in UFO phenomena, often contributing to the mystery surrounding these sightings. While many UFO reports can be attributed to natural phenomena or human-made technology, a significant portion remains shrouded in secrecy, leading some to believe that the true nature of these encounters is being deliberately concealed. This chapter explores the role of government secrecy, military operations, and potential cover-ups in the UFO mystery, particularly focusing on the New Jersey sightings and similar cases worldwide.

The Legacy of Government Secrecy and UFOs

The relationship between government agencies and UFOs has been a subject of intrigue for decades. In the 20th century, numerous UFO sightings were either downplayed, dismissed, or actively covered up by governmental authorities, leading to widespread speculation about what the authorities were really aware of. Many UFO enthusiasts believe that these cover-ups are not only an attempt to protect national security but also to prevent panic and maintain control over information related to extraterrestrial phenomena.

- **Project Blue Book**: One of the most well-known government investigations into UFOs was Project Blue Book, conducted by the United States Air Force from 1952 to 1969. While the official stance of the program was to identify UFOs and determine if they posed a threat to national security, many of the reports collected under this project were either dismissed as misidentifications or left unexplained. The program's closure and the continued lack of transparency led many to believe that the government was withholding crucial information about the true nature of UFOs.

- **The Roswell Incident**: Perhaps the most famous UFO cover-up in history is the Roswell Incident, which took place in 1947 in New Mexico. Initially reported as the crash of a "flying disc," the U.S. military quickly changed the story to describe it as a weather balloon. The conflicting accounts and subsequent secrecy surrounding the incident have fueled speculation that the U.S. government had recovered extraterrestrial technology and beings, only to cover it up.

- **The Role of the CIA and Other Agencies**: The CIA has long been linked to UFO secrecy, particularly through programs like the U-2 spy plane and the A-12 Oxcart, which were developed during the Cold War. These aircraft were so advanced that they could easily have been mistaken for UFOs. The CIA's tendency to dismiss UFO sightings, combined with their control over classified military projects, has led to suspicions that they have been aware of extraterrestrial technology and possibly even extraterrestrial visitors, but have chosen not to disclose this information to the public.

The Influence of Military Operations on UFO Sightings

Military operations, both classified and unclassified, are often at the center of UFO sightings, particularly when they involve advanced aircraft, experimental weapons, or new technologies. While most UFOs reported by civilians are likely to be conventional aircraft or drones, some UFOs may be the result of military operations that are either deliberately concealed or not fully understood by the public.

- **Test Flights of Advanced Aircraft**: Many UFO sightings, particularly in areas with high military activity, can be attributed to the test flights of cutting-edge aircraft that are still in development. The U.S. military, for example, has a history of testing highly classified aircraft, such as the SR-71 Blackbird and the F-117 Nighthawk, which could easily have been mistaken for UFOs due to their unusual appearance, speed, and maneuverability. These aircraft often operate in restricted airspace, away from civilian observation, but accidents or leaks can lead to sightings and speculation.

- **Radar and Technology Interference**: One of the hallmarks of many UFO encounters is the interference with radar systems and other detection technologies. Military aircraft, particularly stealth or experimental craft, are designed to evade radar detection, and in some cases, this has led to sightings of objects that seem to appear and disappear without explanation. Similarly, the use of electronic warfare technology, which jams or disrupts radar signals, can create the illusion of a UFO. Such operations, while classified, are often misinterpreted by civilians who are unaware of the ongoing military activities in their area.

- **The Role of Black Projects**: "Black projects," or unacknowledged military operations, are another area where UFO sightings may arise. These are projects that are so secretive that even those within the military may not be aware of them. Black projects often involve experimental technologies, and their tests can result in sightings that seem to defy conventional explanations. The high level of secrecy surrounding these projects means that the public may not be informed about the true nature of what is being tested, leading to speculation about extraterrestrial involvement.

The Impact of UFO Cover-Ups on Public Perception

The role of government secrecy and military operations has had a profound impact on public perception of UFOs. As more and more sightings have occurred, particularly in areas with high military presence, the theory that governments are actively concealing information about UFOs and extraterrestrial life has gained significant traction. The lack of transparency surrounding these sightings has only fueled the belief that the true nature of UFOs is being deliberately withheld from the public.

- **Misinformation and Disinformation**: One of the primary tools used by governments and military organizations to maintain secrecy is the dissemination of misinformation and disinformation. Misinformation refers to false or misleading information that is presented as fact, while disinformation is deliberately false information intended to deceive. In the case of UFOs, governments have often provided misleading explanations for sightings, such as attributing them to weather phenomena or military tests. In some cases, these explanations are deliberately misleading to keep the public from learning the true nature of UFO phenomena.
- **The Role of the Media**: The media has played a significant role in shaping public perception of UFOs. Throughout the years, media outlets have been complicit in either spreading misinformation or

ignoring the issue altogether. In some cases, the media has been used as a tool to control the narrative and prevent the public from learning about military operations or classified projects that could explain UFO sightings. The failure of the media to investigate or report on these issues has led to further distrust and skepticism among the public.

- **The Rise of UFO Disclosure Movements**: In response to the lack of transparency, UFO disclosure movements have emerged, with individuals and groups advocating for the release of classified information related to UFOs. High-profile figures, including former government officials, have spoken out about their experiences with UFOs and their belief that governments are hiding information about extraterrestrial encounters. These movements have gained significant attention, particularly as more declassified documents and footage of UFOs have emerged in recent years.

UFOs and the New Jersey Sightings: Military and Government Involvement

In New Jersey, as in many other locations, UFO sightings are often associated with military activity, classified operations, and government secrecy. The high density of military installations in the area, combined with frequent reports of strange aerial phenomena, has led some to speculate that these sightings are the result of military or government operations that remain undisclosed.

- **Naval and Air Force Bases**: New Jersey is home to several military facilities, including naval bases, airfields, and research centers. These locations are often the sites of classified tests and experimental operations, some of which may involve advanced aircraft or technologies that could be mistaken for UFOs. The close proximity of military facilities to populated areas increases the likelihood of sightings and subsequent misinterpretations of unusual aerial phenomena.

- **Military Exercises and UFO Sightings**: Many UFO sightings in New Jersey, particularly those involving multiple objects or erratic behavior, may be linked to military exercises. Large-scale drills or tests involving multiple aircraft or drones can easily be mistaken for UFOs, especially if they occur at night or in poor visibility conditions. These exercises may involve classified equipment, which is often withheld from the public to protect national security.

- **The Role of the New Jersey UFO Reports**: As part of the growing interest in UFO sightings, several reports from New Jersey have gained attention for their unusual nature. Some of these reports describe lights in the sky that seem to hover or move in patterns inconsistent with known aircraft. The proximity of military installations in the area, combined with the secrecy surrounding military operations, suggests that some of these sightings may be linked to classified activities. However, the lack of official acknowledgment has led to speculation about cover-ups and the true nature of the sightings.

- **The Simpson prediction:**

The **Simpsons** has long been known for its uncanny ability to predict future events, and one of its more interesting predictions involved a story line about a **fake alien invasion**.

In **Season 9, Episode 1 ("The City of New York vs. Homer Simpson")**, aired in 1997, there's a segment where **Krusty the Clown** is involved in a television special that fakes an alien invasion. The scenario includes elaborate media coverage of the event, manipulating public perception and fear.

Although this isn't a direct prediction of a UFO-related event, it does touch on the concept of **fabricating a major event** for political or media reasons—an idea that some conspiracy theorists and commentators have connected to discussions around potential future **false flag operations** or staged events, including faked alien invasions.

In terms of UFO and alien phenomena, there have been references to the idea of **staging or faking alien invasions**. People have speculated about the possibility of governments or organizations creating such events for political control, economic gain, or to unite populations in times of crisis, something that could tie into this specific **Simpsons** episode's themes.

While the episode itself was satirical, the recent interest and increasing focus on UFOs and extraterrestrial life have made such "predictions" seem more interesting in retrospect, especially when there are ongoing discussions about the **U.S. government's disclosure of UFO information** and conspiracy theories surrounding those events.

Conclusion: Unveiling the Truth Behind Government Secrecy

The role of government secrecy and military operations in UFO sightings cannot be understated. While many UFO sightings can be explained by natural phenomena, drones, or human-made technology, there remains a significant portion of encounters that are likely the result of classified military operations or government cover-ups. In the case of New Jersey and similar locations, the combination of military presence and secretive operations has led to widespread speculation and mistrust. As the push for greater UFO disclosure continues to grow, the hope is that the truth about these phenomena will finally be revealed, putting an end to decades of secrecy and speculation.

11

Drones, Technology, and Human-Made Explanations

While the possibility of extraterrestrial involvement remains one of the most captivating theories surrounding UFO sightings, a growing body of evidence suggests that some of the recent UFO reports, particularly in New Jersey, may have technological explanations rooted in human-made inventions. In recent years, drones, advanced military aircraft, and other cutting-edge technologies have begun to blur the lines between what is known and what is unknown. This chapter delves into the rise of drone technology and other man-made phenomena that could explain many of the UFO sightings observed in New Jersey, and explores how these technologies may lead to future misunderstandings and misidentifications of airborne objects.

The Evolution of Drone Technology

Over the last two decades, drones have transformed from a niche military tool into a widespread technology used in commercial, recreational, and even scientific fields. Modern drones, both unmanned aerial vehicles (UAVs) and unmanned combat aerial vehicles (UCAVs), are now capable of performing complex tasks and can be difficult to distinguish from other aerial phenomena. Understanding the rapid development and capabilities of drone technology is key to explaining many of the recent UFO sightings.

- **The Growth of the Drone Market**: The consumer drone industry has exploded in recent years, with drones now capable of carrying high-definition cameras, sophisticated sensors, and even autonomous flight patterns. Drones from companies like DJI, Parrot, and others are widely used for photography, mapping, agriculture, and even emergency services. However, their increasing prevalence has led to greater confusion when unidentified flying objects appear in the sky.

- **Military and Surveillance Drones**: Beyond commercial drones, the military has significantly advanced drone technology, producing UAVs with impressive capabilities such as stealth, high speeds, and precise maneuvering. Examples like the MQ-9 Reaper and RQ-170 Sentinel are capable of flying at high altitudes with low visibility, and their advanced sensor arrays may allow them to carry out surveillance over vast areas. In addition, these drones may be employed in covert operations that remain undisclosed to the public, leading to speculation that such vehicles could be mistaken for UFOs.

- **Swarming Technology**: Another emerging field in drone technology is swarming, where multiple drones are coordinated to perform complex tasks autonomously. These drones can fly in tight formations, creating dazzling light displays or performing precise movements that may appear otherworldly to observers. Swarms of drones, especially those equipped with advanced LED lights, can easily be mistaken for UFOs, particularly when they exhibit sudden changes in speed or direction.

The Role of Advanced Military Aircraft

While drones have become synonymous with modern UFO reports, advanced military aircraft and classified programs have long been a source of UFO sightings. Historically, many UFO encounters coincided with testing or operations involving experimental military craft, some of which remain highly classified. As technology advances, aircraft are being developed that push the boundaries of what is thought to be possible, making it harder for civilians to differentiate between cutting-edge military technology and extraterrestrial phenomena.

- **Stealth Aircraft and Hyper sonic Flight**: Military aircraft such as the F-22 Raptor and the B-2 Spirit were once considered highly secretive and mysterious when first introduced to the public. These stealth bombers and fighters are designed to evade radar detection, which can make them appear to vanish or behave erratically when seen from the ground. Hyper sonic flight technology, which allows for aircraft to fly at speeds exceeding Mach 5, is also in development, and these aircraft could produce light or visual distortions that resemble UFO activity.

- **The SR-71 Blackbird Legacy**: One of the most famous examples of military aircraft that was often mistaken for a UFO is the SR-71 Blackbird, a reconnaissance aircraft that operated at incredible speeds and altitudes. While not in service today, it paved the way for modern military technologies that could explain some UFO sightings. The aircraft's ability to maneuver rapidly and at high altitudes could have made it appear as a strange object when seen by civilians on the ground.

- **Black Projects and Unacknowledged Aircraft**: There is a long history of so-called "black projects" in the military, referring to highly classified aircraft and technologies that remain hidden from the public.

These projects are often developed with experimental technology that surpasses known civilian aircraft capabilities, and their tests are frequently conducted at high altitudes or in restricted airspace. This secrecy has led to numerous UFO sightings, particularly by those who may have glimpsed a highly advanced, but unrecognized, craft in the sky.

Misidentifications and Optical Illusions

Another significant factor in UFO sightings is the human tendency to misidentify unusual, but explainable, phenomena. The human brain is wired to recognize patterns, and in the absence of clear information, it often resorts to familiar explanations — including the idea of flying objects — to make sense of the unknown. This tendency, combined with atmospheric conditions and technological advances, can lead to numerous misidentifications.

- **Reflections and Atmospheric Phenomena**: Unusual weather conditions, such as temperature inversions, high humidity, or storms, can cause atmospheric phenomena that appear as unidentified flying objects. For example, reflections of lights from distant sources, such as aircraft or city lights, can be distorted by weather patterns, creating the illusion of hovering or moving lights in the sky.
- **Skydivers and Balloons**: High-altitude skydiving and weather balloons can sometimes appear as UFOs when viewed from the ground. Skydivers, especially in groups, can make large, erratic movements that are difficult to interpret from a distance, and weather balloons often float at high altitudes, reflecting light in ways that make them look like objects in motion.
- **Optical Illusions and False Memories**: The human mind can also

be tricked by optical illusions. Certain objects, like aircraft at night or satellites, can appear to be stationary or moving in ways that seem unexplainable, especially when viewed in certain lighting conditions. Moreover, witnesses may misremember or embellish their experiences, adding to the confusion surrounding UFO sightings.

Drone Swarms and Light Displays

One of the most striking features of recent UFO sightings, particularly those reported in New Jersey, has been the appearance of bright, moving lights. These light displays, often characterized by swift changes in direction and speed, are now widely believed to be the result of drone swarms rather than extraterrestrial visitors.

- **Coordinated Light Shows**: Drones equipped with LEDs or light arrays are being used for performances, often in large groups. These light displays can create the illusion of a UFO or strange aerial phenomenon, as drones move together in synchronized formations, changing color and intensity in time with the choreography. The advanced autonomy and precision of modern drone technology allow for such displays to appear almost otherworldly.
- **Drone-Generated UFO Sightings**: Reports of multiple objects moving together in a swarm pattern could indicate the use of coordinated drones, particularly if the objects show erratic or synchronized movements. Drones can be programmed to fly in specific patterns, making them appear as one object or a series of rapidly changing lights, which could easily be misinterpreted as a UFO.
- **Military and Civilian Drones in New Jersey**: New Jersey has been a focal point for drone-related sightings in recent years, with reports of multiple objects flying in formation, hovering in place, and

making rapid maneuvers. These reports are consistent with what would be expected from a military drone test or a civilian drone light show, such as those employed in large-scale events.

The Future of UFO Sightings in a Drone-Dominated World

As drones and advanced technology become more commonplace, UFO sightings may continue to increase, leading to even greater confusion and speculation. Understanding the capabilities of modern drones and military aircraft will be critical to interpreting future UFO reports, particularly in densely populated areas like New Jersey, where aerial activity is likely to be more prevalent.

- **Drone Regulation and Disclosure**: As drone technology continues to advance, the need for stricter regulations and more transparency in government and military drone programs becomes more pressing. This could help reduce confusion surrounding UFO sightings and provide clear answers to the public about the technology being used in their skies.
- **UFOs in the Age of Technology**: The proliferation of drones, satellites, and advanced military craft will continue to challenge our understanding of UFO phenomena. As technology improves, we may find that many of the UFO sightings in New Jersey — and around the world — are the result of human innovation rather than extraterrestrial intervention.

Conclusion: Human-Made Phenomena and UFO Misunderstandings

As we move further into the 21st century, the line between UFOs and human-made technologies is becoming increasingly blurred. While some sightings may still be unexplained, it is likely that many of the reports are the result of drones, advanced military aircraft, and atmospheric phenomena that remain misunderstood by the public. As we continue to investigate these sightings, it is crucial to consider the rapid advancements in technology and how they might explain the phenomena we once attributed to the unknown.

12

Conclusion: The UFO Enigma and Our Search for Answers

As we have explored throughout this book, the phenomenon of UFO sightings, particularly those reported in New Jersey, raises a multitude of questions that challenge our understanding of the world around us. From potential technological advancements to extraterrestrial theories, psychological explanations to government secrecy, the UFO mystery remains a complex and compelling enigma. Each chapter has presented different perspectives, yet the central question—What are UFOs?—remains largely unanswered.

The Persistence of the Mystery

Despite advances in technology, the increasing number of sightings, and the growing number of individuals and institutions studying UFOs, there is still no definitive explanation. The persistence of these sightings—often observed by multiple credible witnesses and recorded under various conditions—suggests that we are dealing with a phenomenon that is far more complex than what can be easily attributed to drones, military aircraft, or natural phenomena.

The Search for Meaning

Throughout the book, we've examined various hypotheses: Are UFOs simply advanced human-made technologies? Are they evidence of extraterrestrial life visiting Earth? Could they be a product of human psychology, mass hysteria, or even altered states of consciousness? While each hypothesis offers a potential explanation, none has been able to fully account for all the factors and characteristics of UFO sightings. What we do know is that the quest for answers requires not only scientific inquiry but also a willingness to entertain new possibilities and expand the boundaries of human understanding.

The Role of Government and Secrecy

Government involvement in the UFO mystery cannot be overlooked. Secrecy, classified documents, and the gradual release of previously withheld information raise further questions about the true nature of the phenomena. Why have governments been hesitant to provide full disclosure, and what are they protecting? These open questions hint at the possibility that UFOs may be linked to areas of research, technology, or encounters that we are not yet prepared to fully comprehend.

The Journey Forward

As we look ahead, it is clear that the UFO mystery is far from being solved. However, each new sighting, every piece of evidence, and each theory contributes to a growing body of knowledge that continues to fuel the conversation. The search for answers is not just about uncovering the truth behind UFOs but also about exploring the limits of our understanding and our place in the universe.

The UFO phenomenon is not just a mystery; it is a window into the unknown. It challenges us to question our assumptions, confront our fears, and expand our horizons. Whether we are dealing with undiscovered technologies, alien visitors, or something entirely beyond our comprehension, the search for answers will continue to shape our collective future.

In the end, the UFO mystery remains an invitation—an invitation to look beyond the ordinary and consider the extraordinary, to explore the unknown, and to remain open to the possibilities that lie just beyond our current grasp.

About the Author – Ignotus

Ignotus, the pen name behind this thought-provoking work, has chosen to remain anonymous, deliberately shrouding their identity in mystery. The decision to write under a pseudonym stems from a deep-seated belief in the power of ideas over personal recognition. Ignotus' primary focus is not on personal fame but on sparking thoughtful discourse and encouraging readers to approach controversial and unexplained phenomena with an open, inquisitive mind.

By remaining anonymous, Ignotus seeks to detach their work from the distractions of personal bias or public perception, allowing the content to speak for itself. This anonymity allows readers to engage with the material without preconceived notions about the author, fostering an environment where the ideas and insights contained within the book take precedence.

Ignotus' decision to remain in the shadows is also a response to the themes explored within the book—concerning mysterious and often unexplained events, including the topic of UFOs and other enigmatic occurrences. The author feels that this approach is symbolic of the very nature of the phenomena explored: often hidden, elusive, and resistant to simple explanations.

This anonymity enhances the sense of mystery surrounding the subject matter and invites readers to consider the true essence of truth and discovery, free from the constraints of identity or expectations. Ignotus' work is a testament to the idea that sometimes the most important questions are not those answered by a known figure but by the collective curiosity of those willing to look beyond the surface.

www.ingramcontent.com/pod-product-compliance
Lightning Source LLC
Chambersburg PA
CBHW070352130626
46556CB00007B/3148